Published by Creative Education
123 South Broad Street, Mankato, Minnesota 56001
Creative Education is an imprint of The Creative Company

Designed by Stephanie Blumenthal
Production Design by Patricia Bickner Linder

Photographs by: Earth Images, FPG International, GeoIMAGERY,
Rainbow, and Tom Stack & Associates

Library of Congress Cataloging-in-Publication Data

Bernardy, Catherine J.
Fuel / by Catherine J. Bernardy
p. cm. — (Let's Investigate)
Includes glossary.
Summary: Briefly discusses the origins of different kinds of
fuels and how they are used.
ISBN 0-88682-989-5
1. Fuel—Juvenile literature.   [1. Fuel.]   I. Title.
II. Series: Let's Investigate (Mankato, Minn.)
TP318.3.B47   1999
662'.6—dc21   98-8291

First edition

2   4   6   8   9   7   5   3

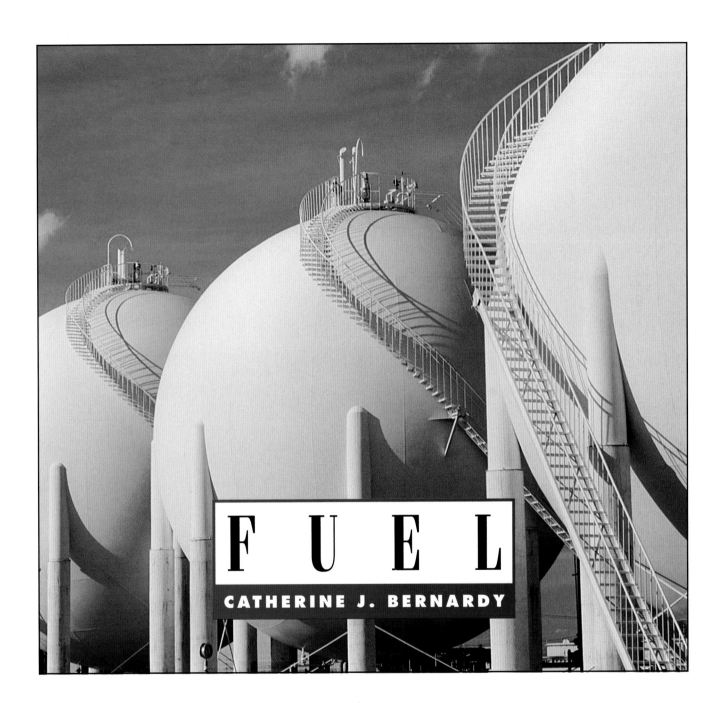

FUEL

CATHERINE J. BERNARDY

Creative Education

FUEL

*Nuclear materials may remain **radioactive** as brief as a millionth of a second or as long as several billion years.*

4

High-voltage towers

Long ago there were no machines. The use of televisions, cars, and gas stoves divides modern times from the past. But all our machines would be a heap of scrap metal if it weren't for one thing— the **fuel** that runs them.

WHAT IS FUEL?

Fuel is anything we use to make **energy,** which is the power used to make things run. Most of the time, we create energy by burning fuels. Long ago, people burned wood and animal dung as fuel for heating and cooking. Today we burn mostly **coal, oil,** and **natural gas** for electricity and heat. These three fuels are called **fossil fuels.**

Unloading wood chips

FUEL
F A C T

*Cars in Brazil run on "neat" fuel, gasoline that is 100 percent **ethanol;** it is safer for the environment.*

FUEL
EFFICIENCY

Ethanol can also be made from straw, cornstalks, sawdust, and even recycled newsprint.

FUEL
SURPRISE

Coal that was formed more than 200 million years ago has been found in Antarctica, which means there must have been water and plants on the icy continent at one time.

Natural energy lab pipes

Some crops, such as corn and soybeans, are used to make vehicle fuel. **Nuclear power** is created by splitting **atoms,** and solar power is created by collecting energy from the sun. Wind power is created when wind blows through a windmill, and water power is created when water spills over a dam. We harness Earth's inner heat to create geothermal power.

HOW FUELS ARE USED

Modern people use fuel 24 hours a day. We use fuel to create **electricity,** which runs the clock radios that wake us up and the lights we turn on in our homes and schools. Some experimental cars even run on electricity!

FUEL
SWEET

A Hawaiian sugar mill is trying out an idea: generating electricity with left-overs from sugarcane processing.

Below, electric car

FUEL
FACT

Geothermal power is generated by steam-driven turbines; the steam is produced by piping water over hot rocks underneath the Earth's surface.

Above, geothermal station, New Zealand; right, an oil refinery in Nova Scotia

Most cars and trucks run on **gasoline** and diesel fuel that we pump into our vehicles at gas stations. Many gas stations sell gasoline which contains ethanol (made from corn) and **biodiesel** fuel (made from soybeans). Rocket ships such as the space shuttle use a special kind of fuel, as do jet planes.

We use natural gas in our ovens and stoves to cook our meals. Natural gas also runs our furnaces in the winter and our hot water heaters year-round so we can take baths and wash dishes and clothes.

9

FUEL
FACT

Out of the estimated eight trillion tons of coal in the world, only 350 billion tons can be mined.

FUEL
FUTURE

Some researchers hope to mine asteroids for minerals and water. Ice under an asteroid's surface could be used to make breathable oxygen in space.

Coal-fueled power plant

Power plants, called utilities, turn the fuel into electricity or gasoline and then sell it to us. Fuel is there for us at the flip of a switch or turn of a dial. If you live in a house or apartment building, chances are you have an electric meter that records how much electricity you and your family use each month.

FUEL
EXPERIMENT

Magnetohydro-dynamics uses a gas or a liquid metal passed through a magnetic field to generate electricity.

Fish fossil

FOSSIL FUELS

Coal, oil, and natural gas are fossil fuels. They provide most of the energy we use. Fossil fuels come from plants and animals that lived millions of years ago. When these plants and animals died, they sank to the bottom of seas or marshes, where they decomposed. Layers of dirt and rock built up over them, creating heat and pressure. This process turned them into coal or oil.

FUEL
FUNNY

Biodiesel fuel made from recycled restaurant fryer grease and burned in the experimental "Veggie Van" makes exhaust that smells like french fries!

Raw coal

Today, we dig mines and drill deep into the Earth to bring coal and oil to the surface, where we can burn it to make energy. Natural gas is given off when we drill for oil. Fossil fuel deposits are found in many places in the world, including the Middle East, Alaska, Asia, North America, and under the oceans.

Ancient Middle Eastern people first discovered natural gas seeps between 6000 and 2000 B.C. Other ancient peoples used crude oil and asphalt more than 5,000 years ago. The Chinese used coal to melt and refine copper as early as 1100 B.C. They also were the first to drill for natural gas in 211 B.C., a fuel unknown in Europe until 1659.

Coal mining

FUEL

POSSIBILITY

Someday we may be able to use liquid nitrogen in cars. The process of liquefying nitrogen removes carbon dioxide from the air; this will reverse pollution.

Donkey pump extracting crude oil near a vineyard

Oil-soaked fibers provided the ancient Persians with flaming arrows during the siege of Athens, Greece, in 480 B.C. The peoples native to North and South America used oil as medicine, and the ancient Egyptians used it as a lotion to soothe wounds.

Left, coal-fired cogeneration plant; below, an electric meter

CHANGING NEEDS

For centuries, fossil fuels were used sparingly. Then, during the 1800s, the **Industrial Revolution** changed the way people used fuels. Europeans and Americans invented many machines that operated on large amounts of fuel.

FUEL

OCEAN

Scientists around the world are trying to harness power from ocean tides to generate electricity.

FUEL

TRASH

In some cities, electric power plants are fueled by garbage.

16

Right, smokestacks turn out pollution; far right, an oil rig in the North Sea

Factories that made everything from furniture to clothing sprang up in Europe and North America. All of these factories used fossil fuels for power. The railroad industry's steam engines needed a lot of coal. At the beginning of the 20th century, Henry Ford began to mass-produce the automobile. The oil industry prospered because gasoline is made from oil.

FUEL
SCARE

In 1874, Pennsylvania's state geologist warned that the U.S. had only enough petroleum to last four years.

18

Within decades, the world had been transformed by machinery and methods of transportation that used large amounts of fossil fuels. People marveled at such new inventions as electric washers and dryers, elevators, and tractors.

Not only does technology need fuel, but fuel needs technology. Oil companies use computers to map the inside of the Earth and find natural gas and oil deposits.

The Alaska pipeline

FUEL
F A C T

More than half the world's supply of crude oil—700 billion barrels— is located in the Middle East.

They inject air and water into wells to force oil upward, retrieving what couldn't be reached before. Improved processes also mean more gasoline and diesel fuel squeezed out of every barrel of oil. Cheap fuel is the basis of the world's **economy.** Low fuel prices mean companies can sell their products for less money, leaving people with more money to purchase more of those products.

FUEL
DANGER

Coal miners who breathe in coal dust over long periods of time are prone to a dangerous illness known as black lung disease.

Burning old tires creates pollution

POLLUTION

Have you ever seen a picture of Los Angeles when the wind isn't blowing? The air isn't hazy and brown because of fog or clouds—it is full of **pollution.**

Pollution forms when fossil fuels are burned and **chemicals** are pumped into the air. But what goes up must come down. When it rains, these chemicals fall back to the earth in the form of acid rain, acid snow, or acid fog.

A cid rain makes water undrink-able; it kills birds and fish, harms crops, and ruins buildings. Pollution can also happen in the ocean when companies drill for oil. Sometimes oil pipes break or ships carrying oil have accidents. When this happens, oil leaks into the water, killing fish, or it washes up on shore, killing wildlife.

FUEL
CONCERN

Temperatures over the past decade have been the hottest on record for the Earth, probably due in some part to the burning of fossil fuels.

Above, an oil spill off the coast of Spain

FUEL
PRODUCTS

Chemicals derived from oil and gas are used in making such products as cosmetics, detergents, medicines, and fertilizers.

An oil spill off the coast of France

Cars and trucks also make pollution. When we drive, our cars produce foul-smelling exhaust that contains two harmful gases. **Carbon monoxide** is a deadly gas that can kill people who breathe too much of it. **Carbon dioxide** doesn't kill, but it traps heat near the Earth's surface. Scientists around the world now believe that this trapped heat is raising the Earth's temperature. This change is called **global warming.**

Between 70 and 80 percent of France's electricity is generated from nuclear plants; the country has the lowest electric costs in Europe.

A raised temperature around the world could affect the weather, causing more hurricanes, flooding, and droughts.

Global warming and the limited supply of fossil fuels have worried scientists for decades, leading researchers to develop other fuel sources, such as nuclear power and fuel crops that can be grown.

Trucks are used to deliver oil

24

A nuclear power plant and nearby wildlife ponds in Oregon

NUCLEAR POWER

Nuclear power is a type of energy created when atoms are split. The chemical reaction that happens as a result produces huge amounts of energy. This energy heats water, which produces steam when it boils. The steam is used to generate electricity.

At one time, people thought nuclear power could meet most of the world's energy needs. A tiny amount of nuclear fuel can produce enormous amounts of energy. Some "breeder" nuclear reactors even produce more fuel than they use.

There is a problem with nuclear fuel, though. After it is used, the waste is radioactive. It can make people and animals sick, and even kill them. Engineers encase nuclear waste in lead and bury it, but it can be radioactive for millions of years. No one has developed a good way to get rid of these **toxic** wastes.

FUEL
SAVINGS

Trees that shade houses can reduce the amount of energy used for air conditioning by up to 50 percent.

FUEL
STUFF

Byproducts from refining oil are used in making asphalt, candle wax, and furniture polish.

FUEL
FACT

The country of Iran has triple the natural gas reserves of the United States.

Right, Chernobyl plant disaster, Ukraine; far right, Spring Hill nuclear plant, Pennsylvania

People also fear nuclear accidents. In 1979, an accident happened at Three Mile Island in Pennsylvania, and another occurred in 1986, in a town called Chernobyl in the former Soviet Union. The accidents released harmful levels of radiation into the environment.

Many people oppose nuclear power, so its development as a fuel source has been slow in the U.S. Also, nuclear reactors are very expensive to build.

FUEL
B L E N D

The U.S. government has determined that a 10-percent ethanol blend gasoline for cars will reduce carbon monoxide emissions by 25 percent.

A logging operation in California

GROWING FUEL

Fossil fuels are not **renewable** because they will eventually run out. In the U.S., some coal power plants cut down on pollution with the use of crops. They burn biomass (plant products such as leaves, corn stalks, and bark) as well as coal. Even sawdust and newspapers can be burned as fuel.

FUEL
MISTAKE

A one-billion-dollar nuclear plant in France sat idle for many years and produced only six months of electricity; it was finally shut down in 1998.

Ethanol is made of corn

Cars and trucks using renewable fuels produce less pollution also. The corn product, ethanol, is added to the gasoline we use for our cars. Biodiesel, made from soybeans, is a fuel for trucks and buses. These fuels burn more cleanly than gasoline, and thus cut pollution. There is no threat that corn and soybeans will run out. Farmers raise them year after year.

FUEL

SAFETY

Biodiesel burns at a lower temperature than standard diesel fuel, making it safer to use than diesel.

Top right, wind generators; bottom right, solar photoelectric panels; far right, the space shuttle Discovery

Renewable fuels help reduce pollution and slow down global warming. They also benefit a nation's economy. Farmers can produce these fuels to sell to local power plants. These fuels do not have to be mined or drilled from the ground, nor must they be imported from other countries.

Technology has changed the world quickly. We have many fuels and fuel uses today that we did not have 100 or even 50 years ago. Also, life is much different now than it was at the beginning of the century. It is certain that fuel use in our lives will change much in the future as well.

Glossary

Atoms are tiny particles, invisible to the eye, that make up everything in the universe.

Biodiesel fuel is made from soybeans and is used in buses, trucks, and other vehicles that burn diesel fuel.

When fossil fuels are burned, **carbon dioxide** gas is given off. This gas traps heat in Earth's atmosphere.

Carbon monoxide is a colorless and odorless deadly gas given off when fossil fuels are burned.

Coal is a solid fossil fuel that is mined from the earth.

Chemicals are substances out of which all things are made.

The **economy** is the production, trading, and use of goods and services of a country.

Electricity is a fundamental energy that can provide power to run engines and turn lights on.

Energy is the power or ability to do work.

A fuel made from corn or other plant starch that is used to run vehicles is **ethanol.**

Fossil fuels are the fuels formed from fossilized plants and animals that died millions of years ago. Fossils are the preserved remains of these plants and animals.

Any substance that can be used to create heat or power is a **fuel;** coal, natural gas, and wood are fuels.

A fuel that is blended from natural gas and oil is **gasoline;** we use gasoline in cars.

Global warming is a rise in the Earth's average temperature. It is also called the "greenhouse effect."

The **Industrial Revolution** was a period in history when people first began to invent and use machines and fuel.

Natural gas is a fossil fuel found alongside coal deposits.

Energy made from the force of splitting atoms is **nuclear power;** splitting atoms is called fission.

Oil is a liquid fossil fuel used to make gasoline, kerosene, and many other products.

Waste products in the air, soil, and water that harm the environment create **pollution.**

Radioactive is the term used to describe something that gives off radiation. Exposure to radiation is harmful to people, animals, and plants.

Anything that can be easily replaced is considered **renewable.**

Anything harmful or deadly to life is **toxic.**

Zero-emission from burning a fuel means that no harmful waste products are given off.